Middlebury Historical Society

The marble border of western New England

Its geology and marble development in the present century

Middlebury Historical Society

The marble border of western New England
Its geology and marble development in the present century

ISBN/EAN: 9783742828613

Manufactured in Europe, USA, Canada, Australia, Japa

Cover: Foto ©Klaus-Uwe Gerhardt /pixelio.de

Manufactured and distributed by brebook publishing software
(www.brebook.com)

Middlebury Historical Society

The marble border of western New England

THE MARBLE BORDER

OF

WESTERN NEW ENGLAND

MIDDLEBURY HISTORICAL SOCIETY,

MIDDLEBURY, VT.

MARBLE BORDER

WESTERN NEW ENGLAND.

ITS GEOLOGY AND MARBLE DEVELOPMENT
IN THE PRESENT CENTURY

PAPERS AND PROCEEDINGS OF THE

MIDDLEBURY HISTORICAL SOCIETY.

VOL. I. PART II.

CONTENTS.

MIDDLEBURY HISTORICAL SOCIETY.

THE MIDDLEBURY HISTORICAL SOCIETY was organized November 17, 1843, as an association having in view, as a specialty, the collection of New England and American history, and the promotion of an interest in that department of intelligence in the community. There were at first but nine gentlemen associated, which number was only once increased, by election, ... all residents of Middlebury, the plan of action being mainly that of a literary club. Its officers were a president and secretary, the latter also treasurer and curator of collections. The officers first chosen were Hon. Samuel Swift, president, and Philip Battell, Esq., secretary, who were continued in office until 1868, when Rev. Benjamin Labaree, D. D., was chosen president, and George S. Swift, Esq., secretary. A standing committee of two members besides the secretary was constituted for all general purposes not otherwise provided for. New members were elected from time to time from the town and vicinity, honorary members in Addison County, and corresponding members out of it. Meetings were held monthly, or quarterly, as occasion required, at which historical papers were read and discussions held, the society with some collections of manuscripts and pamphlets being preserved. In February, 1859, a project was adopted to procure the preparation of town histories of all the towns in Addison County, with a view to their publication as soon as prepared. A special committee was constituted for this object, by whom historians were selected in each town, and commissioned by a circular from the society formally defining the comprehensive work to be done. The appointments were accepted in almost every case and an interest in the design was awakened in the towns. Eight or ten of such histories were prepared and reported. Of these, four in a complete form and of high merit were published, chiefly by the towns concerned and by local subscriptions. The first was that of Middlebury, with a preliminary historical account of Addison County, introductory to the series, by Hon. Samuel Swift, in 1859. The History of Salisbury, by John M. Weeks, Esq., followed in 1860; that of Shoreham, by Rev. Josiah F. Goodhue, in 1861; that of Cornwall, by Rev. Lyman Matthews, in 1862. Histories of Orwell and Bristol were also prepared, by Hon. Roswell Bottum and Hon. Harvey Munsill, of which the former was published in 1880. The society has continued since 1843 an annual celebration of the landing of the Pilgrims, regarding the Plymouth colony as the starting point of New England history and its character as typical of American civilization. The plan of organization of the Society is still maintained. In 1843 it was the nineteenth of the historical associations in the United States, as recognized by the Department of Education in 1876, when the number had increased to eighty, many of them representing with distinction their relation to the country and acting as active conductors upon others in the interests they serve. The present officers of the Society are:

HON. JOHN W. STEWART, ... *President.*
PHILIP BATTELL, Esq., ... *Secretary.*
PROF. EZRA BRAINERD, } *Standing Committee.*
RUFUS WAINWRIGHT, Esq., }

TRODUCTORY.

[illegible] *Register*, an article subsequently reprinted, to which a brief appendix, first read to the Society, was also published.

In the meantime the society had been struck with the character and extent of the investigations which had thus incidentally been opened to it. Whatever the mechanical form might have been, it was evident that their chronology had been but little disturbed; that while art or artistic [illegible] had struggled along with some [illegible] [illegible] of [illegible] from one generation of people to another, from one art into another, with glimpses of [illegible] public and private upon it, art in history which [illegible] the elements of social progress in man, if possible with spirit infused, was indifferent to it. The more comprehensive and spirited the modern book, [illegible] the [illegible] the [illegible] that [illegible] its inadequacy. The more intelligent the general student whom a correspondent might consult, the more startling his confession that the subject was quite new to him, and except in the case of one [illegible] [illegible] and [illegible] of investigators in New England, with the aid accessible to them it seemed likely to remain so. The research had [illegible] enough to look for the [illegible] if its [illegible] [illegible] were hidden, and trace it to its origin.

From this point, illustrated by Professor [illegible] of Yale College, the [illegible] of the course of American [illegible] history was in the [illegible], and the [illegible] of it followed up the [illegible] [illegible] by the [illegible] [illegible] through the [illegible]. To these the clues at some [illegible] [illegible] were already in hand, for happily the relationships of society within the [illegible] border once so familiar were not entirely obsolete, and the fellowship of [illegible] studies, [illegible] by [illegible] [illegible] interest in the subject, facilities of inquiry were [illegible] [illegible] [illegible] of [illegible] [illegible] long periods ago. The town histories disclosed something, [illegible] in some instances as to [illegible] in the trade, the [illegible] [illegible] of the [illegible] [illegible]. [illegible] to those in their [illegible], who had not forgotten the fact. But the living sympathy of man face to face is the truest interpreter between them, and by that impressions slowly or partially rendered were [illegible] or extended.

Professors Brainerd and Seely had been particularly interested in carrying out the investigation, the former representing the department of applied science in Middlebury College, the latter that of Chemistry and Mineralogy. The entire suite of the principal localities, at present or formerly operated, was visited by them south of Middlebury during their April vacation, and these were reported at the regular meeting of that month by a display of specimens obtained in the excursion, and from localities north, all distinctly characteristic, illustrated by their colloquial explanations and a sketch of the border map prepared by Mr. Brainerd. The completed papers of these gentlemen were read at the regular meeting June 19th, when it was voted that the society shall enter upon the design of the publication of its collections with the timely and interesting articles now read; that they be referred to the Standing Committee for publication, together with the correspondence elicited by the investigation and other related matter. The official statement, prepared by Professor Seely, of the relevance of the local question of ownership, with the decision of the Reference thereto is also given in the appendix.

The geographical traits of the New England border have been controlling in its history. As the country reaches upon the Sound it has its character with that of the tide-water system, along which the first settlements extended. But soon another jurisdiction was met at the east and in receding from the coast a profile of boundary corresponding to it. The character of the surface at the east favored the institution of a mere mastery between these States, and the movement of population obeyed it. The historian sees it in masses from Stratford and Milford, movements acting successfully upon it, encircling and itself in separating gradually in the continuous boundary the new parishes and towns in the Woodbury territory and New Milford beyond. The border is more distinctly defined at the north. Massachusetts has its Berkshire county, which extends uniformly across the State, and has never been without a history or witnesses in the better spheres of life as much as known. It rests upon the kindred commonwealth of

lege of President Dwight and a contemporary in service in the revolution. Perhaps from a higher field of observation than as an observer merely, he writes to General Philip Schuyler of New York, alluding to a conversation had between them the winter before in Philadelphia, and suggesting the resources of Vermont which might contribute to sustain a canal proposed to be built between the Hudson river and Lake Champlain. "There are also," he says, "in this part of the Country numerous quarries of marble"—the letter is dated at Rutland, January 15, 1792—"some of them of superior quality. Machines may easily be erected for sawing it into slabs by water, and in that case it might become an important article of commerce."

The Otter river from Middlebury having worn the marble barrier away, that once set it back upon its longest and broadest level stretch, forms a winding channel of descent as by terraces of similar material, six miles, to the Weybridge cataract, afterward in a broader stream, the finest interior river of the State, to the wide triple fall of forty feet at Vergennes, where it reaches the Champlain level. Dr. Dwight was attended here by another former pupil, Col. Seth Storrs, the earliest resident lawyer at Middlebury, and found Judge Enoch Woodbridge, first mayor of the city in 1794, from his circle of friends at Stockbridge, Mass. He notices the city charter of the place, its commercial chances, but chiefly the views shown him from the cupola of the court house, the mountain ranges on either hand so newly grand to him; the lake system having closed off the Vermont line begun so far to the north, and opened upon the then nameless mountain wilderness of Northern New York, the land border to the north skirting the "glass," as he elsewhere calls the lake at Burlington, and widening as the Green Mountains recede.

P. BATTELL,
E. BRAINERD,
R. WAINWRIGHT,
Standing Committee.

THE MARBLE BORDER.

ITS GEOLOGY AND DEVELOPMENT.

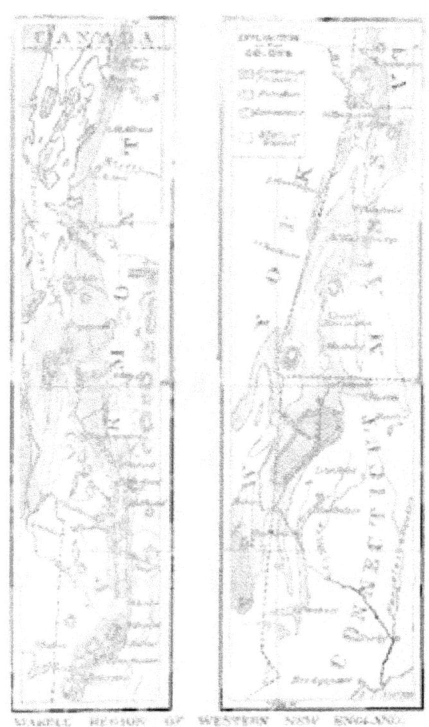

MARBLE REGION OF WESTERN NEW ENGLAND

THE GEOLOGICAL FEATURES

OF THE

MARBLE BELT.

BY PROF. EZRA BRAINERD.

THE marbles of Western New England in variety of color, in fineness of texture and in durability surpass those of any other region of the United States. They are found chiefly along the western borders of the States of Connecticut, Massachusetts and Vermont, between the Green Mountain uplands which extends from the Canada line nearly to Long Island Sound, and the ... Taconic mountains which extend south of Lake Champlain and in places extend the marble zone within the border of New York. The rivers which drain this valley are the Housatonic flowing south to the Sound, the Hoosic and the Battenkill flowing west as affluents to the Hudson, and the Otter Creek flowing north to Lake Champlain, the first and last contained wholly within it. The object of this paper is to present briefly the geological views which have obtained in respect to the marble district and to indicate those more recently established.

The pioneer explorer in American geology was Professor Amos Eaton. He was a leader also in other departments of science; his popular "Manual of Botany," Dr. Asa Gray says, "was for a long time the only general work available for students in this country." In 1817 Prof. Eaton delivered his first course of public lectures to the students of Williams College, from which he had himself graduated in 1799. His success was so great that he determined to continue giving courses of lectures on the natural sciences. This he did in many places—among others to the Medical school in Castleton, Vt., to the students of Middlebury College in 1817–18, and to the members of the New York Legislature at Albany. Here he found success—had the good fortune to find a munificent patron in Gen. Stephen Van Rensselaer, who sustained him for nineteen years in his geological labors, enabled him to travel more than 15,000 miles in explorations through New York and neighboring States, published his maps and reports, and finally established the Rensselaer Institute at Troy, with Mr. Eaton as its senior professor.

Prof. Eaton's views in regard to the rocks of Western New England were for many years the accepted views. In 1820 the science of geology was in its infancy even in the old world. It was [illegible] by theories, and the [illegible] theories of Werner [illegible] this respect a characteristic [illegible]. The fact, that each formation had its peculiar group of fossils, that marked the formation the world over, and distinguished it from every other formation, was not understood. The only method Prof. Eaton knew of, to any extent, for determining the relative age of strata was to trace them separately through a vast extent of country and to compare their general characters.

According to Mr. Eaton's system, there are five series of rocks in the earth's crust: I. The Primitive; II. The Transition; III. The Lower Secondary; IV. The Upper Secondary; V. The Tertiary. Each series is subdivided into three formations: The Carboniferous, the Quartzose and the Calcareous. The five-fold recurrence of these three formations marks the structure, [illegible] to system and regularity; though it hardly conforms to the complex facts of nature. Only the two oldest series of rocks were found in our territory,—the Primitive and the Transition. These were distinguished mainly by the fact that the former was destitute of any organic remains; the latter contained [illegible] of molluscs and remains of [illegible] on the earth. Prof. Eaton calls all the rocks of Western New England Primitive, excepting the [illegible] western border of Vermont and Massachusetts. The line of separation passes through Middlebury, Rutland and Williamstown, and from the southwest corner of Massachusetts trends toward the work across the Hudson just below Poughkeepsie, and on through Orange County to New Jersey. It will be seen, then, that all the granite of Western New England falls into the Primitive series and is, in fact, the upper formation of that series.

Prof. Eaton supposes that these earlier formations were covered the whole earth like the concentric layers of an onion; but that under them along the different north and south lines combustibles were placed of [illegible]. At the close of the Transition Age, these combustibles were ignited and produced a great explosion, which uplifted the mountain chains of the world. There have been, subsequently, several explosions; but it is supposed that now the combustible materials are too nearly exhausted to produce anything more than ordinary earthquakes.

There was much to admire in the character of Professor Eaton. His pupils, among whom we find [illegible] and Hall, were warmly attached to him, and inspired with his own ardent enthusiasm for science. He possessed an imagination that was bold and vigorous in conceiving [illegible] as a scientist. [illegible] that theories must be constantly subjected to revision and modified or set aside to [illegible] with newly discovered facts, and such was Prof. Eaton's habit of mind. He realized the peculiar uncertainty attending geological speculations; confessed that he had been obliged by [illegible] to make so many changes in his successive books, as to provoke the charge of fickleness. But he claimed to be ready to abandon any of his views, if further investigation should render them untenable. He understood [illegible] what he was leaving the science of geology; and in one of his last communications to the "Journal of Science," in 1839, [illegible] with fresh, hopeful and [illegible] geology [illegible] seven thousand pupils or auditors," had something to do with it. At any rate, at his death, in 1842, every State in the Union, except five, had authorized geological surveys. Massachusetts was the first to publish a report, ably prepared by Edward Hitchcock, afterwards president of Amherst college. Dr. Percival made careful explorations in Connecticut, while the survey of New York was entrusted to four eminent geologists, Hall, Emmons, Mather and Vanuxem, and its thoroughness of execution surpassed all others. Owing to the wide range of formations in that State and their wealth of fossils, these reports have long been standard authorities in American geology. The true order of the various formations along Lake Champlain was determined; their characteristic fossils were identified; and Potsdam sandstone followed by Calciferous, Chazy and Trenton limestones and these by Hudson River slate remain today in geological science substantially as reported in 1842.

Not so, however, with the rocks lying directly east of the Champlain group. Over these arose a memorable controversy which can not even yet be said to have wholly ended. Dr Emmons, long professor at Williamstown, in his volume of the final New York report, advanced his theory of the "Taconic System." He claimed that the range of mountains extending from Addison County in Vermont south along the western borders of Massachusetts and Connecticut, and also the limestones and marble on the east of the range, belonged to a formation older than the Potsdam but younger than the Primitive rocks. In this opinion he was opposed by Professors Hall and Mather, his colleagues in the New York survey, and by Prof. Rogers of the Pennsylvania survey. These regarded the limestones and slates of the Taconic range as belonging to the Champlain group, only altered by heat and more intensely disturbed.

Meanwhile the State of Vermont tardily awoke to the importance of a geological survey of her territory. It was first entrusted to Prof. Adams of Middlebury College, who made several annual reports. But when he left the State, in 1847, the work was left for several years in an unfinished state. It was at length put into the hands of Prof.

and a formation of great thickness, which he named the Quebec group; (3) that a great fault or crack existed in these strata, in which the rocks on the east were uplifted from one to two thousand feet and thrust up over on to the rocks to the west. This fault passes along the western base of Snake Mountain, Sneb Mountain, Philo Mountain, Lone Rock Point near Burlington, and thence along the western shores of the lake into Canada. Then, bending more and more toward the east, it passes through Quebec just north of the fortress and on down the St. Lawrence to Point Gaspe.

These conclusions have stood the test of subsequent research, and may be regarded as established facts in American geology. But Sir William Logan went farther. After an examination of the Green Mountain region, he extended his Quebec group in 1868 so as to cover the whole territory. In fact, he pushed its conquest onward through southeastern New York into New Jersey and Pennsylvania, and also eastward into Maine and New Hampshire. This was indeed only a milder form of the Taconic theory, the only difference being that it came in just above the Potsdam and calciferous formations, instead of just below. It was a broad generalization, just as sweeping and almost as arbitrary as that of Dr. Emmons. But for a while there was no one bold enough or sufficiently equipped with facts to dispute the views of so high an authority as Sir William Logan.

To investigate them was the task of a modest, obscure Vermonter, the Rev. Augustus Wing. He was a graduate of Amherst College in grade in the school house or the abandoned country church. In 1869 he laid down for himself the task of determining the age of the marble formation in Western Vermont. "Knowing," says Prof. Dana, "that fossils were the only sure criterion of geological age, he searched and he found them, and thus reached safe conclusions." "His investigations made more for the elucidation of the age of Vermont rocks than had been done by the Vermont Geological Survey." "His discoveries shed light not on these rocks alone, but also on the general geology of New England and Eastern North America."

But while of preeminence in the field, he was slow in making known his discoveries. His excessive diffidence and aversion to writing had much to do with this; his desire to get further facts and to make his work complete and exhaustive perhaps had more. It was only by a letter from Mr. Billings of Montreal, to whom Mr. Wing had sent fossils for identification, that Prof. Dana first learned in 1872 of the importance of these discoveries. He wrote Mr. Wing for a "fuller account of the fossils of West Rutland." Mr. Wing wrote only two pages in reply, but never finished the letter nor sent it. Prof. Dana visited him in August, 1875, and obtained from him a promise to send to the "Journal of Science" an account of his discoveries. But the article was never sent, and Mr. Wing died the following January.

His note books and unfinished letters were afterward sent to Prof. Dana, who compiled an account of his discoveries for the "Journal of Science" in 1877 (pp. 332 and 405, vol. xiii). While we are left in ignorance of many important details and of the precise location of many of his fossils, his general conclusion is established beyond all doubt. It is that the limestone formation of Western New England containing the marble is the same as the Calciferous, Chazy and Trenton of the Champlain group, and that the slates of the Taconic range overlie the limestones and belong to the Hudson River and the Utica formations of the New York Reports. The Georgia slates of Franklin County are not paralleled with the slates of Rutland and Bennington Counties, Vt., and of Berkshire County, Mass., as Prof. Emmons maintained and as the Vermont Report supposes. The Cacrude system of Emmons finally disappears from American geology; while the Quebec of Logan is reduced to a subordinate member of the limestone group, if its existence is to be recognized at all in Western New England.

It only remains to notice the confirmations that Mr. Wing's conclusions have since received. Trenton, Chazy and Calciferous fossils have within the last few years been discovered in the northern portion of the Green Mountain area, in the supposed Taconic slates and limestones, and this in many localities by Profs. Dana, Dwight and Whitfield. Prof. Dana has furthermore spent much time during the past seven years in a thorough and detailed examination of the rocks of Western Connecticut and Massachusetts and of Southern Vermont. Though these rocks are much disturbed and metamorphosed, so as to be destitute of fossils, Prof. Dana's labors nevertheless place beyond all question the age of the limestone formation in these regions, and tend to show that the whole Taconic Mountain range belongs to the same age as the fossiliferous rocks along the borders of Lake Champlain. Portions of older, or "Archean," rocks are believed to occur here and there—reefs and islands in the ancient Silurian sea. Great difficulty attends the detailed elucidation of the central range of the Green Mountains, but we trust Prof. Dana will live to complete his labors. At any rate the age of the marble and the slate has been settled formica slate or into gneiss, according to the degree of heat to which it is subjected.

It is believed, then, that once, in the remote past, the whole Green Mountain area from the Lower St. Lawrence to the present site of New York City was a gulf or arm of the ocean. At first the tides or currents of the sea were such as to cause extensive deposits of sand over the greater part of the gulf. But certain limited regions were favorable to the growth of shell fish where calcareous deposits were made. The mottled marble of Northwestern Vermont represents such a region. The quartz of this period is known as the Potsdam sandstone, one of the best exposures of which is at the Ausable Chasm in Clinton County, N. Y.

Afterward, from some oscillation in the earth's crust and some change in the direction or character of the ocean currents, the waters became more clear and the conditions were more favorable for the growth of lime-producing animals. The sea swarmed with various species of mollusca, crinoids and trilobites, and especially with organisms akin to the corals that are now building coral reefs in the South seas. The calcareous remains of these animals formed beds over the entire region of Western New England varying in thickness from 1,000 to 2,000 feet. This may appear incredible; but it should be remembered that the time was indefinitely long.

This second period was brought to a close and the third initiated by the waters of the sea becoming for some reason turbid. The ocean currents may have flowed across a gradually sinking barrier-reef which existed to the east of the Green Mountain area. The waters deposited their fine silt or sand over the immense beds of shells, and the animals that produced them became extinct. The deposition of fine sand and clay continued for a long period.

Then came the era of mountain-building. The stages in the formation of the Green Mountains are the same as those of other mountain ranges. The earth's crust is but a comparatively thin shell (perhaps 30 to 40 miles thick) over a white-hot, if not molten, interior. At a depth of 5,000 feet the average heat is enough to boil water; at a depth of six and one-half miles, enough to melt lead. The natural results of the accumulation of sediments several thousand feet deep along a region where the earth's crust is at the same time gradually sinking in the form of a trough 500 miles long by 50 wide, is, that the internal heat will melt off or soften the underside of the sinking trough, until the region becomes a comparatively weak line in the earth's crust.

At the same time there is a lateral push from the east. For, as the earth is constantly becoming smaller from cooling, the external shell must contract or wrinkle to fit it. The wrinkles naturally occur in the weakest places. By a powerful shove, from the direction of the ocean, the horizontal sedimentary beds are tilted up at a steep angle, crumpled together, folded over one another, and thrust high up above the level of the sea. These movements of the rocks under this enormous strain are a source of heat. The bright light of meteors is known to result from the friction of these bodies against the air. So it is proved by exact calculation and experiment that the crushing and shearing of a rocky stratum may raise its temperature to a red-heat. But a less temperature combined with enormous pressure and the agency of super-heated steam would suffice to convert the sediment into the existing rocks. Such, if we may believe the geologist, were the earth-throes by which the Green Mountains were brought forth.

The events thus far described extended over an exceedingly long period and about the same amount of time has since elapsed. During this latter period the Green Mountains are supposed to have held their heads above the waters of the ocean. But even the eternal hills wear away during the lapse of ages. The rain and the frost, the erosion of rivers and the grinding of glaciers have removed a large portion of the original mountains. Wherever the limestone appears it is certain that the entire slate formation that overlay it has been carried off; and where the red sandstone comes to the surface, both the overlying slate and the limestone have disappeared. A glance at our map, or an acquaintance with the topography of Western New

England, shows that the limestone is usually found in the valleys near the water courses; while the higher peaks and ridges are invariably slate, quartzite or gneiss. The valleys are, then, valleys of erosion, by which the underlying limestone has been exposed.

Attention should be called to the fact that the same limestone appears different in different places. Along Lake Champlain, the more ancient and stable rock of the Adirondack region upheld the sedimentary deposits so that they were less crumpled and less baked by the thrust from the east. These rocks accordingly contain abundant fossils; while those to the east and south are nearly destitute of fossils. On the west the slates are clayey and crumbling; but they become harder and more crystalline as one journeys eastward or southward. The gray sandstone west of Lake Champlain becomes the chocolate sandstone of Western Vermont. The [illegible] limestones of the lake region are gray or black from the carbonaceous remains of the shell-fish. But on the eastern borders of the Vermont area the same beds are often snowy white, and become more and more coarsely crystalline to the south. At the higher temperature to which the latter rocks have been subjected, the black carbon has been burnt out, just as the fire of the lime-kiln often converts a dark limestone into the whitest of lime.

The process of crystallization, moreover, may be compared to the analogous process of tempering steel. In both cases by a certain amount of temperature and subsequent cooling the molecules are grouped in regular crystals, which can be seen with the microscope and often in the case of marble with the unaided eye.

At the same time the fact remains to be accounted for that in the same region but a comparatively small portion of the lime rock has been converted into valuable marble. It is not possible to draw the line sharply between a limestone and a marble. Webster's definition of the latter is "a limestone, hard, either when polished or otherwise, for ornamental uses." The best marble results only from a happy concurrence of many circumstances. A limestone may fail to attain the rank of a marble for one or more of the following

1. It may contain impurities which impart a dull or muddy color. The range of allowable colors is wide; but it must be bright and agreeable to the eye.

2. It may contain minerals that are much harder or softer than the lime-carbonate. Silica or sandy matter is found in most of the lime-rock, rendering it difficult to cut and polish. Occasionally great beauty of color may compensate for this, as in the case of the [illegible] marble, which is quite siliceous.

3. The coarseness of the texture may prevent a pure and beautiful limestone from ranking as an ornamental marble. This is generally the case with the deposits now profitably wrought in Berkshire and Litchfield Counties of Massachusetts and Connecticut for architectural purposes. On the other hand, from a lack of crystalline structure the stone may not be susceptible to a good polish. Rutland and Addison Counties are the region where the crystalline structure is at about the right extent.

4. The unsoundness of the rock may be a grave practical objection. A jointed structure, or the presence of minute fractures, or a tendency to fracture, is a frequent cause of failure.

5. Lastly, a stone, otherwise excellent, may fail in durability on exposure to the weather, and so be unsuitable for many of the uses to which ornamental marble is put. The crystals that compose it are sometimes too feebly cemented together and the stone crumbles or furnishes an easy foothold for lichens. Many impurities, not otherwise objectionable, dissolve away after a while, producing a stain, or, existing as separate crystals, become detached, leaving a cavity. Or sometimes the lime-carbonate itself dissolves away faster than the impurities and the stone "weathers" to a dull or dingy color.

Of the limestone formations of limestone in Western New England (2000 to 3000 feet in thickness), the lower portion known to geologists as the Calciferous (300 to 400 feet in thickness), is for the most part siliceous, partaking of the nature of the sand-rock that underlies it. The upper portion, known as the Trenton (300 to 600 feet in thickness), is impure from the presence of clayey matter, marking of the terms of the slate formation that overlies it. Only certain layers of the middle portion seem to have been fitted by their original constitution for the production of marble. These strata in Rutland and Addison Counties appear in two parallel lines, about two miles apart, stretching from the North line of Middlebury to the South line of Rutland, and are from one to two hundred feet in thickness. This deposit bids fair to remain in future the chief source of ornamental marble for the continent.

The accompanying map is compiled from the geological surveys of the States of Vermont and Massachusetts and from Professor Dana's Manual of Geology.

Split Rock Mountain on the west shore of Lake Champlain represents the peculiar rocks of the Adirondack region. A quarry has been recently opened in this mountain and the stone is known as "Champlain granite." But it is not a true granite. It consists chiefly of black pyroxene and lime-feldspar. The rock belongs to the "Archæan Age," the oldest recognized in geology. Professor Dana considers the area surrounded by the limestone and sandstone in Northwestern Connecticut to be of this Age.

The portion of the map colored red is where the rocks of the Potsdam period come to the surface. The blue shading indicates the outcropping of the limestone and marble of the Canadian and Trenton periods. The uncolored portion of the map, except the vicinity of Troy, indicates the quartzite, the gneiss and the talcose slates of the latter part of the Trenton period.

THE MARBLE FIELDS

AND

MARBLE INDUSTRY OF WESTERN NEW ENGLAND.

BY PROF. HENRY M. SEELY.

THE belt of marble already mapped and sketched, so interesting on account of its geological character, has still other claims to our attention. For nearly a century it has been the seat of an industry, which, in these later years, is becoming one of vast importance, and the region is marked by a line of active or suspended quarries throughout almost its entire extent.

Our study over the field from north to south, a stop at some of the most important localities, inquiries about others, and additional information from various sources have put us in possession of facts which seem to be worth recording and to which I now wish to call your attention.

A rapid glance at the present condition of the industry will prepare us for further historic and statistical details of the work.

Beginning at Middlebury, near the extreme northern limit of the belt of white marble, in the northwest part of the town will be found a quarry which, with its mill and finishing shops, has been in operation the past year. This property, known as the Cutter Marble quarry, has passed through many vicissitudes and through many hands. Theodore Phelps built the first mill with an undershot water wheel and a single gang of saws. David Ralph succeeded him and did a prosperous business in the making and sale of thresholds, window caps, and grave stones. Following Ralph were Dr. E. Vale and Abel Spenc-

To one unacquainted with the working of Vermont marble, the Sutherland Falls quarry opens like a revelation. This wonderful sight occurs, that of a hollow cube cut into a hill with perpendicular walls on the north and west rising to the height of near one hundred feet, open to the sky, and with an acre at part forming its horizontal marble floor. Over this floor are running channeling machines, cutting deep, narrow grooves into the solid stone. Long parallel blocks of stone are thus made; these are cut across into more nearly cubical forms, loosened from their bed, and lifted from the quarry, to be wrought up here at the Falls or elsewhere into finished marble.

The mills have great capacity. Sixty-two gangs of saws are in operation night and day, and additional buildings are being erected to meet the increasing demand for the stock marble.

Finishing shops are in active operation, putting into final shape both architectural and monumental stone.

The various departments of labor at the Falls give employment altogether to about five hundred workmen.

West Rutland, with its many quarries, has given, these later years, the reputation to Vermont marble that has made it famous. While the marble of Sutherland Falls is classed as mottled, much of that quarried at West Rutland is pure white. The strata have here been followed at a steep dip and in one quarry to the depth of two hundred and twenty feet. The abyss made by the huge long slit into the under earth, when filled with ascending smoke and steam from the engines at work below, seems [illegible]. This great opening is partly covered over by the overhanging rock; in the winter it is sometimes further protected by a wooden roof.

Post [illegible] around it, is a mountain of marble, the top bearing a cap of slate. Here in different localities are situated several quarries, worked mostly in the summer time. Early in April, as we passed, the work of the year was not sufficiently advanced for satisfactory observation, so the locality was reserved for future study.

The remaining towns south, in Bennington county, offer no present inducement for delay, and we pass over the border of Vermont into Berkshire County, Mass. North Adams near the Northern line of Massachusetts was early noted for its beautiful white marble. The quarry near, just at an elevation of 250 feet above the village, is no longer worked for marble; the stone of the easterly opening [illegible], however, is selected and ground into marble dust.

Still south on the marble belt are the towns, Lanesborough, Pittsfield and Lenox, but in these no quarry is in operation.

At Lee extensive quarrying for architectural marble is going forward, a force of upwards of sixty men being employed in the work. A mill with four gangs of saws is in operation, but most of the stone is shipped to Philadelphia in the block.

The towns, West Stockbridge, Alford and Egremont, are in the range and at former times have produced marble. None now is raised. Near the village of Great Barrington a [illegible] stone called [illegible], which is worthy of more than local note, is taken out in moderate quantities for architectural purposes.

Further south, is the town of Sheffield, which at one time had a reputation for its enormous blocks of marble, which with great labor were carted across the hills to the Hudson river, whence they were shipped to Philadelphia. This industry ceased at this point years town rich in historic interest, which is still holding its marble quarries, though they have been abandoned thirty years.

Further to the South and West the line of Calcareous rocks runs on into New York, but it is not the present purpose to follow it beyond the borders of New England.

Having taken a hasty run over the region and a glance at the present condition of the industry, it may be proper to return more deliberately over the ground and notice on the way some matters in connection with the history and progress of the work, something about methods, improvements, [illegible], and [illegible], and some facts bearing relation to the geology of the region.

Historically, Marbledale, in the town of New Milford, Conn., has a right to claim first mention. Here near the date 1800 Philo Tomlinson was at work quarrying marble and sawing it into slabs. He had come up the Housatonic valley from Derby in the same State. One may still find at the spot where he located his first mill the remnants of machinery and among them the decaying apparatus by which sand and water were distributed to the saw similar to that for which he secured a patent.

Orcutt's history of the town of New Milford fixes the date of early operations pretty clearly, and the correctness of this date is further established by the headstones in the graveyard at New Milford. The oldest of these are mostly of sandstone from the Connecticut Valley, made picturesque, but almost illegible by the covering of lichens. Those of slate, of nearly the same age, hold their lettering much better. The earliest marble headstones were evidently those which had been separated from the rock in the quarry by hand drilling and wedging, and then worked down by hand. But those bearing a date soon after the beginning of the century were cut on one or on both sides by a saw.

Although future search may carry further back the date of the invention of the various pieces of machinery useful in working marble, to Tomlinson of Marbledale must be accorded the credit of combining these in a serviceable form, and applying them successfully to his purpose in sawing marble. The saws were toothless saws, stretched in a frame forming a gang, the gangs were driven by water-power, and the saws were fed by sand and water, equally distributed by an automatic arrangement.

The saws used by marble workers in the time of Pliny and centuries before was a toothless saw fed by sand and water. But the saw was drawn by hand.

Water power was used in Germany for sawing stone as early as the fourth century, but this mode of sawing seems to have fallen into disuse and to have been forgotten. Near the year 1730 William Colles of Kilkenny, Ireland, restored and put into use the practice of sawing marble by water power, and near at Kilkenny it has been continuously in use down to the present day.

The Irish chronicler, William Tighe, in [illegible], noticing this industry in Kilkenny, states: "The principal work is done at the marble mill which is on the left bank of the river (Nore) near two miles from Kilkenny. This mill is admirable for its simplicity." The same machinery drove both saws and polishers. An account of this mill was published in London in the year 1748 in which the writer of a "Tour of Ireland," states: "I cannot learn that any one has attempted to imitate this machinery. It is perpetually at work, by night as well as by day, and requires but little attendance."

The saws here used were of soft iron, and were at first and for a long time often of single blades, not working in a gang.

A knowledge of the process apparently went over to Derbyshire, England, for in the year the "Tour of Ireland" was published, works similar to the Irish were established. The English authority, Tomlinson, in his book on the Useful Arts, says: "The principal marble manufactures of England is in Derbyshire, along the valley of the Derwent and the Wye, from below Buxton to Derby. The machinery for sawing and polishing was first established at the village of Ashford, near Bakewell, in 1748, water being the motive power. About 1800 similar machinery was erected at Bakewell. Both works are situated

XVII., p. 343.

Yet with the greatest information it does not seem to have been in general use until after Philo Tomlinson's application of it at his mill at Marbledale.

The industry flourished at Marbledale where it originated. Quarry after quarry was opened and mill after mill erected on the banks of a small stream, the (East) Aspetuck. In 18[illegible] as many as fifteen quarries were actively worked, and a like number or more mills were in operation within a distance of less than three miles, and the marble went to all parts of the country.

Soon after this date competition from other localities, especially from Dover, N. Y., and Rutland Vt., began to cause a decline in the profits of the work. This competition increased, particularly from Vermont, the work became unremunerative, and now also the Goodsell quarry, one of the principal—and last in operation—was abandoned. The sawing likewise declined until at present a single mill at the village of New Preston, doing local work, is the only existing representative of the industry.

Why the work, so well established, was compelled to give way to competition may not be plain at first view, but the cause will be readily explained on the examination of the rock in place. The strata at Goodsell's quarry are of a steep dip, not quite perpendicular, while the beds affording the best marble are not thick. Much unproductive stone must be removed, rendering quarrying expensive. The rock is not a pure calcareous marble, but magnesian, and through it some minerals like tremolite and asbestos are found. The white rocks not far away from the quarries are breaking down into a white sandy mass, showing that the material does not weather well. To the other ness of the rock, not yielding massive blocks, and the lack of enduring quality may be largely charged the failure of the industry.

In the northern part of Litchfield County near the Massachusetts line is the town of Canaan, and in East Canaan is the quarry that furnished the marble for the new State House at Hartford. The rock lies in massive beds, near the surface, and is apparently in vast quantity. The rock is white and coarsely crystalline, contains in some portions of it many crystals of tremolite, and though not suitable for monumental uses, is fitted for the purpose of building to which it has been put. Large blocks can be obtained, and one may expect that the work which has been stopped will be revived. Derricks are still standing, indicating the recent abandonment of the quarry. The remains of the large steam mill may be seen; the mill itself, however, has been burnt to the ground.

The saws used in this mill were peculiar in this, that they were toothed with diamonds. The diamonds were set in pieces of steel or iron, and these pieces were bound into the blade of the saw. It is reported that they were wonderfully effective, cutting the blocks of marble very rapidly.

The explanation is made that the East Canaan quarry closed work because the contract for which it was opened being completed, there arose at that time no further demand for marble of this quality.

Sheffield, Mass., adjoins Canaan, Conn., and has had several quarries in operation. The quarry in the north part of the town, from which the large columns for Girard college, Philadelphia, were obtained (1832–1846), was noted for its large blocks, and from these blocks sometimes fifty feet long were blasted out. The use of gunpowder in removing blocks from the quarry is not often commendable, but with the care exercised in this case the operation seems to have been quite satisfactory.

The same marble range extends through the eastern part of Egremont and Alford, but the quarries in these towns, like those of Sheffield, have been abandoned.

A quarry of clouded marble was at one time, 1847, opened in the town of Great Barrington, but is not now worked. But in two or three localities near the village there are quarries of the *blue stone* previously mentioned, which, if found to weather well, ought to find favor among architects. It has been used in the construction of a public building at Pittsfield, with excellent effect. The new church and rectory in Great Barrington are of this stone, as are other structures, as the Berkshire Hotel, built years ago. There is a warm and cheerful look about these buildings, such as does not characterize those constructed of white marble. The character is largely a crystalline magnesian limestone, but it carries with it many grains of quartz and iron pyrites. The grains of pyrites are mostly microscopic and compact, though in some places the quantity is so great as to be notably evident. This ingredient may interfere with its usefulness, but on [illegible] no decay was observed in the structures which we hastily examined.

Stockbridge, West Stockbridge and Lee lie north of Great Barrington and are on the range, though the limestone is in some places invaded by slate, apparently dividing the rock into various fields. The quarries at West Stockbridge and Lee are those which are most worthy of note, the former for its early history, the latter for its present products.

[illegible] Ames, Newell & Clark, as early as 1800 erected a stone saw mill at the "old quarry" below West Stockbridge village. It is well established that the same persons had a mill in the village certainly as early as 1803, and with great probability a year or two earlier; so the industry of sawing marble here, as well as at Middlebury, began at the beginning of the century.

In November, 1803, Johnson & Stevens concluded a contract for furnishing the marble for building the front of the New York City Hall (1803–1812), having early in the year bought a tract of land on which was opened a quarry, later known as the Fitch quarry, which by a change of town lines is now in the town of Alford. Under this contract over 33,000 cubic feet were delivered at a price of $1.06 per foot. Later, that is, in 1808, the same firm filled another contract of [illegible] a foot, the stone at this time probably coming from the old quarry.

How valuable the stone was for building is shown by the fact that after an exposure of more than three-quarters of a century it is practically without sign of crumbling or decay. The extension of the capitol at Washington is from marble obtained at West Stockbridge. The quality and beauty of the marble, useful for monumental as well as architectural purposes, brought it a most desirable reputation and orders for it came from all parts of the country, East, South and West.

Berkshire County produced in 1845, according to the statistics of Noel Dewey, marble to the value of [illegible]. The product of 18[illegible] Mr. Chas. B. Boynton estimated to be $100,000.

The demand for the West Stockbridge marble was much ahead of the supply. Under this stimulus the mills were kept running night and day. Later, steam was tried in sawing, but on account of the expense, which was asserted to be [illegible] cents per foot, against [illegible] cents a foot by water power, it was abandoned.

Chas. B. Boynton introduced several appliances which at the time were thought to be very helpful for quarrying and dressing the marble here. He revived the process invented by Wiseman, and used more than a century before in the marble quarries of the Pyrenees, of cutting the blocks from the rock by saws. Wiseman used saws over twenty feet long. Boynton's saws are not described, but he had the advantage of his long-time predecessor in having steam as his motive power, which he used in driving his saws.

ity was required, but also straight mouldings and grooves could be cut with great facility and exactness.

At one time seven sawmills were in active operation in West Stockbridge. Fortunes, in those days considered large, were accumulated in the business.

The rock in its best layers is nearly a pure calcium carbonate. These layers dip, as do most of the marble strata, to the East. The old quarry pitches under the West Stockbridge River and has been worked for some distance under its bed.

Quarrying stopped here near 1855. Fereday's mill is still standing. Blocks of marble are piled upon each other by the side track of the railroad, waiting for transportation. But the machinery of the mill has gone into decay, while the decrepit derrick would break its spine in attempting to raise a block to a car.

The cause of this stopping of the work was sought for. Major Robbins (aged 83), who saw the quarry and mill in its time, who witnessed the rise, growth, maturity, decline and cessation of the industry (1800–1850), he himself having had a large part in the work, gave tersely a full answer to our inquiry by saying, "Quarrying stopped at West Stockbridge because Italian and Vermont marbles were better and cheaper."

The quarry at Lee has escaped the abandonment that has fallen upon those of other towns in Massachusetts. The life of the quarry is evinced by the stroke of the drill, the puff of the steam engine and the sweep of the channeling machines. The thick strata are quarried under the open sky from a nearly horizontal floor. Six channelling machines are at work preparing the rock to be loosened in blocks. A diamond "gadder" is also used. Steam is used in the mill for sawing as well as in the quarry for driving the channelling machines.

The stone is not suitable for mortal or monumental purposes, but is quarried chiefly for buildings. Philadelphia, a city demanding much marble for its public edifices, furnishes the largest market for the product of Lee's quarry. The large blocks are often pieced with mono-

[illegible]—dog tooth, in the language of the quarrymen—resembling that of the stone from East Canaan, Conn.; while the stone is not pure calcium carbonate, but a dolomite. Lime is burned near the quarry, but the stone is not obtained from the same beds as those furnishing marble.

The towns of Lenox, Pittsfield, Cheshire, Lanesborough, New Ashford, Adams and Williamstown, all have a share in the marble belt and some of these early shared in the prosperity of the industry. Particularly was this the case with Pittsfield, New Ashford and Lanesborough. From Lanesborough was obtained the marble used in the construction of the old capitol building at Albany. The rock is represented to be of fine grain and good quality. Lack of convenience for transportation has been suggested as interfering with the successful working of the beds.

The rock at North Adams, though of good character chemically, is too coarsely crystalline, as before stated, for a good quality of monumental marble, while the beds are not massive enough to give large blocks for architectural purposes. The rock is perhaps best recognized by the use to which it is now put, that of pulverization, by which a beautiful marble dust is produced.

Including East Canaan quarry close adjoining Berkshire, from which the handsomest state capitol in New England was constructed at Hartford, 1878, that of Sheffield, furnishing the columns of Girard College, 1833, the material for the extension of the capitol at Washington, 1865, and for the City Hall now building at Philadelphia, West Stockbridge leading all others with the New York City hall, 1803–12, the most conspicuous public structures have been supplied from the deposits in Berkshire county, neither of which in any sense has been exhausted. Other industries may compare for its capital, employing cheaper labor, but none strengthens a claim of gratitude to its seat of learning or do not lend majesty to recount its valley.

Approaching the localities in which the business of quarrying marble has centered these last years we come to Dorset, in Bennington County, Vermont. The quarries in this town are situated mostly up the sides of Dorset Mountain or Mount Aeolus.

It is stated, with much truth, that this mountain is a mountain of marble. But it must not be inferred that it stands out as a pure block of marble and can be cut into anywhere and commercially quarried. Many of the strata are siliceous and of too hard a nature to be worked, others are slaty and fragile, some are magnesian, and some are ferruginous, while in some places layers of quartzite and in others of slate occur through it. Comparatively, but a small part of the mountain furnishes merchantable stone, and yet the small proportion is in the aggregate enormous.

As one ascends the mountain he finds most of the rock obscured by debris, making it difficult to observe with accuracy the real sequence of the different strata.

He will pass beds of variable character, such as the compact and fine grained marble, higher find the beautiful white and more coarsely crystalline stone, still higher, the gray interbanding with white, then the gray carrying many crystals of pyrite and films of hydromica slate, then he will miss the limestone altogether, which disappears under a debris of contemporaneous rock, over which have fallen many blocks of slate. Higher still are masses of slate, mixed in many places with native quartz and by this slate the top of the mountain is covered.

This cap of slate Hitchcock estimates to be 998 feet thick and the limestone with the marble below 2,500, the whole a total of 3,498 feet for this wonderful mountain. The actual junction of the slate with the limestone, here, as at almost every point throughout the whole marble region, is obscured by debris.

sition of this rock and the adjoining slates are also the deposits of iron ore and kaolin, which so generally border the slate belt and attend the deposits of marble.

Lower down, after passing the brecciated white and grey limestones, is reached the snowy, evenly-grained marble, coarsely crystalline and white. Flag-like layers here afford slabs suitable in thickness for headstones and the early supply of marble here came from such strata. Deeds for such quarrying property date back before the present century. From one hundred and fifty to two hundred feet below the slate the white architectural marbles occur in, and over four hundred feet lower the fine-grained marbles useful for ornamental and cemetery purposes are quarried. Below this the rocks are usually less suitable in character to yield valuable marble.

The strata here, although they do not have a high dip to the east, as is almost always the case at marble quarries along this line, lie nearly horizontal or gently undulating.

The first quarry opened in Dorset—following the authority of Mr. F. Field in the Vermont Gazetteer, [illegible]—was by Isaac Underhill, in the year 1785, on lands then owned by [illegible] and still held by the [illegible] family. The quarry was first wrought for fire jambs, chimney backs, hearths and lintels for the capacious fireplaces of the day. People came a hundred miles for these beautiful fire-place stones and considerable trade in them soon sprang up. Other quarries were soon in operation and from 1785 to [illegible] nine quarries were opened. The first marble gravestone ever finished in Dorset, it is believed, was the work of [illegible], in [illegible], from a slab taken from the [illegible] quarry. Active manufacture of marble headstones seems to date from [illegible], when Elijah Sykes opened his quarry and gave his chief attention to this business. The stock was taken from the top or outer edge of the layers where the strata could be readily split into sheets or flags of a thickness of four inches or so. These sheets were then cut down with chisel and [illegible] to the desired shape.

Marble was first sawn in the town of South Dorset in [illegible] by [illegible] and Brown. When this method of sawing was well established the former sources of inferior marble were abandoned and better stone used. This stone marble found a market in New York, Boston, Philadelphia, Cleveland and at intermediate points. In 1840, before the introduction of Italian and Rutland marble, the demand for Dorset marble was beyond the supply.

Various quarries are in operation upon the mountain. At South Dorset fine monumental stock is obtained at the quarry of Kent & Root. Channeling machines and rock drills with a force of perhaps sixty men are at work. A mill with eight gangs of saws is in operation.

S. F. Prentiss & Co. are working a quarry higher up the mountain with a force of about thirty men. From this a beautiful building stone, both of white and bluish shades is obtained. A part of this stone is shipped directly to Philadelphia in blocks, part sawed at the mill not far away and another part at their mill at East Dorset. The annual product may be set down at from 20,000 to 30,000 cubic feet.

Up the mountain at East Dorset, the Dorset Marble Company is working what is known as the Freeman Italian marble quarry. The stock here produced is used almost exclusively for monumental and decorative purposes. Some of the beds, of which as many as twenty are recognized, afford very beautiful mottled and clouded marble. The quarry was worked forty years and more ago. Masses of rock of immense size are moved or thrown down by a blast and then worked into manageable blocks by hand or machine drilling. The blocks are carted down to the mill, where twelve gangs of saws cut the stone into desired shapes. An annual average of 30,000 cubic feet is produced.

the dip varying from 25° to 85° with an average of 45°. The thickness of the beds wrought is from 90 to 110 feet and these have been followed to a depth of 200 feet. The strata have many marked and distinct layers, varying in color, texture and value. White, clouded and blue are the prevailing marbles; the white, fine-grained, statuary marble being the most valuable.

Full statistics are not at hand indicating the number of quarries and the force employed. Approximately the number of quarries may be set down at fifteen, and the number of men engaged about the quarries, mills and shops at two thousand.

The methods of quarrying at West Rutland are like those in other places and will be described further on. Steam is the power used in the mills as well as in the quarries.

The ingenious device for automatically distributing sand and water to the saws in the marble mills, which was invented by Philo Tomlinson of Middlebury and used in the mills there the first half of the century, seems to have fallen out of usage. In most of the mills one man is still required to watch two gangs of saws, keeping up the supply of sand and caring for its proper distribution. Recently other methods have been proposed. One just introduced at West Rutland is a simple elevator arrangement of pans and buckets. By means of this one man will be sufficient to superintend six or eight gangs.

A successful application of the "sand blast" as cutting means for ornamental purposes was made at Sheldon & Slason's works at West Rutland, about the years 1873–1876, in preparing headstones for soldiers' graves at the national cemeteries. Similar work has since been done at other localities, but on a smaller scale.

The blast of sand upon marble rapidly wears it away, but if a portion of the surface be protected from the abrading action of the sand, that part will soon stand out in relief. By using the ingenious process perfected by Tilghman the work was rapidly and cheaply done. The parts of the stone to remain raised were protected by an iron shield, while letters and figures of chilled iron were placed upon the spaces where the inscription was to be raised. The blast of sand, issued on to the prepared stone would eat away the unprotected part, while that protected would rise up in high relief out of the block. By this method, the name, company, regiment and rank of a soldier could be put upon a stone in less than five minutes of time. Contracts were filled by which [illegible] lettered headstones, having dimensions of three feet in length, ten inches in width and four inches in thickness, were placed in the national cemeteries at no expense to the government of $[illegible]. The economy of the process alone made it possible to place these monuments at the soldiers' graves for so small an amount.

The stone has been thought too soft for statuary purposes, though some layers have been used with great satisfaction for this purpose. At the State House at Montpelier may be seen the statue of Ethan Allen of heroic size. This statue was cut by Larkin G. Mead, and appropriately from Vermont marble, the quarry at West Rutland furnishing the stock. While, from its softness, the marble from these quarries cannot carry so fine lines as the Italian, still it stands exposure to the weather better. The layers, too, are growing harder as they are followed into the earth, and some of them are about as hard as the foreign.

Before reaching Sutherland Falls, still in the town of Rutland, there is to be found the quarry of the Columbian Marble Company. The marble here is of a peculiar character, variegated and almost black. This has obtained an enviable reputation for interior and ornamental work, and from its firmness and hardness stands at the head of the list of dark marbles.

The extent of the quarry at Sutherland Falls has been already alluded to. The marble is of the clouded variety. Some beds opened in this quarry are much like those that farther south furnish the Columbian marble and farther north the Mountain dark.

It would be interesting to know the origin of these mottled marbles in which the white and black are so curiously mingled. The material giving the black appearance is seemingly carbon, and probably in character approaching the graphite which marks the [illegible]

power is admirably economized. The saws in the mills, the polishing beds and other machinery of the finishing shops and even the drills in the quarry are driven by it. For work in the quarry the air is compressed by water-power and conducted in an 8-inch pipe about half a mile away. There would be no gain in using compressed air in an open well-ventilated quarry if steam were necessary for the compressors. Really there would be a loss of power amounting to about 40 per cent. But here, as the falls afford so much surplus power, compressed air is exceedingly economical and effective.

The mechanical appliances at work at the quarry used at Sutherland Falls are worthy of notice. Some of these have already been incidentally alluded to. It is by the introduction of channelers and power drills that the profitable working of inclined ledges of rock is entirely secured.

An early view of marble quarrying would give us the sight of a few men, perhaps spared for a little time from the labor of the farm, working away with drills and wedges at the outcropping and easily separated layers of marble. With these crude, almost rustic, means, the local market would be supplied.

Twenty-five years ago the glimpse of quarrying at West Rutland was given: "It is an interesting scene to behold two hundred quarrymen ranged in rows, each with his long, sharp drill, steadily cutting deeper and deeper those grooves that are destined to sunder the [illegible] from their bed those valuable blocks so much prized here. The musical ring of the quarrymen's drill, first reverberating to the ear, from the deep-sunken quarry, is pleasing to the spectator, as he stands and looks down into it; and to the proprietor it is the welcome harbinger of the 'good time coming,' for a successfully managed quarry is quite sure to bring fortunes to the proprietors."

To-day the view of quarrying is entirely different. The statement of profit may very nearly be as true as when the above was written, a quarter of a century ago, but the description of the method will not fit. The introduction of power has changed the aspect of the quarry, the manual drill being now used only as necessary. The preparations for removing the blocks from the quarry are mostly made by machinery. The clamp of channeling machines and the noise of steam engines driving the diamond drills are harmonious only to the designer. What, however, has been lost in music by the passing away of the old method is more than made up in the efficiency and economy of the new.

These aids to the quarryman may be properly mentioned at this place, since Sutherland Falls has seemed to be the pioneer point for cutting machines.

The channeling machine manufactured by the Steam Stone Cutter Company of Rutland, Vt., was invented by George J. Wardwell of Rutland. The first successful machine was built by him in 1863, in connection with the Sutherland Falls Marble Company, and that original machine has been at work there constantly until within a few months. These machines are now in operation in almost all of the important quarries in the country, and it is calculated that over 3,000,000 square feet have been cut by them. The channeler is essentially a locomotive machine driven by power, usually steam, moving over a steel rail track which is placed on the quarry bed, and carries a single gang drill on one side, or two such drills, one on either side. These drills are raised and dropped by a lever and crank arrangement. The gang of cutters forming the drill is composed of five steel bars, seven to fourteen feet long, sharpened at the ends, and securely clamped together. Of the five cutters, two have diagonal edges; the other three have their edges transverse. The centre of the middle, longest, extends lowest, so that the five form something like a stepped arrangement away from the centre. The drill, lifted, drops with great force, and rapidly cuts a channel into the rock. The single machine is operated by two men, the double by three. As it runs backward and forward over the track the machine is reversed without stopping, and as it goes the cutters deliver their strokes, it is claimed, at the rate of one hundred and fifty per minute. The machine feeds forward on the track half an inch at every stroke, cutting half an inch or more every time of passing. The single machine will

as real Indian, are blending into an industrious and intelligent commu-nity.

Pittsford has the honor of having one of the earliest quarries in Vermont, perhaps the very earliest. Jeremiah Sheldon worked mar-ble here in 1795, taking the material from the quarry now known as the [illegible] quarry. For many years marble for [illegible] for the country around was obtained from Pittsford. The purple calcareous [illegible] would do excellently for the [illegible] and sides, but when used for the back and jambs of the old-time fire-place would fail. A variety of stone from this place, probably magnesian in character, was early sought for and widely used.

There are three beds or veins of marble running through the town north and south. The most easterly of these has a breadth of about one foot and on it was opened in 1871 the quarry known as the Central Vermont quarry. It was successful until the great depression of 1893, from which it has not as yet rallied. The marble is of the same character as that of Florence, or Sutherland Falls, of which bed it is probably a continuation. The middle bed is separated from the easterly one by about [illegible] of lime rock. The bed is of [illegible] feet wide and contains marble of different shades varying from a pure white to a dark blue. The marble is of excellent quality and very sound. Buildings made of this marble have stood the test of years without showing signs of weathering. An illustration of the en-during character of the marble from this bed is seen in the [illegible] at the Court house in Brandon. The Pittsford Quarry com-pany formerly, operating on this, furnished marble for several large buildings in Boston, among which may be mentioned the Continental Building, Commonwealth Hotel and the Blackstone National Bank building.

[illegible] border and are on the outcrop of the same bed. These quarries are known by the name Florence and the station at the railroad has come to bear the same name. The Florence quarries are about two and a half miles north-west from the Pittsford station and about three-fourths of a mile west of the railroad. The supply of Florence marble seems inexhaustible and from its superior quality is fast coming into favor. The [illegible] or west bed, thought to correspond to that of West Rutland, is about half a mile west of the eastern and abuts on the west against the slate, the slate having disappeared from all the region east of it. This bed, about one [illegible] wide, holds beautiful dark blue marble veined and mottled. Such marble has only recently been demanded in trade. Within a year these quarries have been opened to supply the new call for dark marbles. That opened by F. W. Smith & Company is known as the Florentine quarry and has been in great [illegible] almost from the moment of opening.

During the year seven quarries were operated. Considering the extent of the deposits and the character of the marble, no one can [illegible] the bounds of the prosperity of the marble industry in the town of Pittsford. The output of merchantable marble during the year 1884 was nearly 10,000 tons.

Brandon has many varieties of rocks in the marble range, and early after the settlement of the town men were engaged in the production of marble. As many as ten quarries have been opened within the limits of the town, though few of these have produced desirable stone. Of those which have been wrought with considerable activity, beginning with the oldest, is the [illegible] quarry about two miles north-west of the village, closed about [illegible] years ago; the Prout quarry two miles north of the village, not worked; the Billings quarry, between these mentioned, active from about 1825 to 1879. The quarry recently opened (1881), known first as the [illegible] quarry, now called the [illegible], north of the village, is the one in most active operation, employing perhaps fifteen men at the quarry and [illegible] at the mill. This quarry is most admirably situated for convenience, the blocks are cut out from the hillside, placed on tracks and transported easily to the and which is but a few rods away. The stone is [illegible] for monumental purposes and the company working the quarry being large building contractors, use most of the stock for their own work. The other quarries opened, either from the injudicious method of extracting, or from the character of the stone, have not found extensive sales and are for the present abandoned.

The history of the marble industry in Middlebury reaches back almost to the very beginning of the century. According to Prof. Frederick Hall, marble was discovered here in [illegible] by Dr. Ebenezer W. Judd and first sawed by him here in [illegible]. The mining that year was preliminary to a larger business, which was entered upon in [illegible] and continued at the same place until [illegible]. During those years the marble was quarried on the east bank of the Otter just below the falls. It was sawed by water power close by the place where it was obtained. It found a market in New York, Boston and other cities. The amount of sawn marble produced annually in those early days of the business was from [illegible] to [illegible] feet, the standard of thickness being two inches. In [illegible], which year may be taken as a fair average year, the sales amounted to [illegible], the expenses about [illegible]. Some of the marble stones were taken from the bed of the river just above the falls at times of low water.

Other quarries, mostly in the northern and western parts of the town, have been opened. One of the oldest with its mill was early known as the Spaulding or the Parker quarry. It has since been known by various names, as the North Middlebury, Old Middlebury, and Cutter Quarry. By the last name it is now known and is the only quarry in operation in the bounds of Middlebury. Other quarries have been designated as the White or Pure Marble quarry, the Vermont, and the Addison County. In all these, expensive excavations have been made, but they have failed to be remunerative. In some cases beautiful white marble has been obtained, but the amount sold from them and much sawed has been so small that the proprietors have not felt encouraged to continue the work.

It is a matter of prime importance that the [illegible] of this character tion should be determined. If the beds were shattered throughout the whole structure at the change from limestone to marble, no hope can be entertained of successful work on these strata. If, however, the flaws are only superficial and caused by external agency, then deeper excavation will afford sound marble. If the latter is the case, it may be asserted with entire confidence that untold wealth is buried here, waiting for those already convinced, or whom it may invite, to come and uncover it. The opening of large beds of sound rock similar in character to that already obtained and scarcely tested will give Middlebury the reputation of producing the toughest, whitest, best statuary marble of the whole range.

The assertion that the character of the marble of this range deteriorates in grain in going south from Rutland and in soundness in going north, is not well established. It is true that going north some of the layers do become finer in texture, but until the layers deep in the earth have been examined it will be too early to say that throughout the region they are unsound. In grain and texture the white marble of Middlebury is known to be the perfection of beauty. In the deep layers it may be proved to be the perfection of soundness.

Brief mention should be made of such deposits that in some cases have been used as marble. Those bordering Lake Champlain have undergone but little change and so retain, more nearly their original character of limestone.

At Larrabee's Point in Shoreham, Addison County, are deposits of black marble rivaling the finest of Kilkenny in character. Fossil shells appear in some parts, giving a variegated appearance to the stone. At one time a quarry here was wrought with great success. Elaborate and beautifully-polished chimney pieces wrought from this black marble are to be found in some of our older houses. The quarry is awaiting a return of public demand for the dark marbles, when it will yield, as formerly, an excellent material for useful and ornamental work.

been worked out for this purpose is abandoned. But on the south and west end near the west shore of the Island, a quarry, lower down in its geological position, has [illegible] been worked with great success. The massive dark stone is considerably used for ornamental purposes, but is largely [illegible] as [illegible] to be [illegible] in demand in connection with the white marble. Fossil shells of great beauty are not unfrequently found even in the darkest specimens.

Swanton, in Franklin county, formerly furnished a dove-colored marble which was extensively used for gravestones. From this on for many years marble was taken at this point, but since [illegible] Rutland marble displaced [illegible] superseded by [illegible] and [illegible] the quarrying of this beautiful stone was wholly abandoned.

The line of the State of Vermont does not limit the range of marble rock; but it continues into the Dominion of Canada, where, just north of the line, the black marble has been extensively quarried.

A peculiar rock, scattered in boulders above the marble of Addison County, is found in place not far from the shore of Lake Champlain in Addison, Chittenden and Franklin counties in Vermont. This rock is in its geological position just below the true marbles belonging to the Potsdam; but in several localities it has by a series of uplifts been thrown up through the true marble-bearing rocks. The rocks of these up-throws, often called the red sand rock, are coarse in texture where they appear in Addison county, but in the northern part, especially at Mallett's Bay in Chittenden County and near St. Albans, Franklin County, they become so fine [illegible] in character that they are worked as true marbles, and the product is known in commerce agate, jasper, opal, carnelian, and at times serpentine veined in pure white marble. As a fancy marble it cannot be surpassed.

In composition it is a magnesian limestone containing a notable per cent. of silica, iron and alumina. Its texture is very fine and it takes a remarkable polish. One peculiarity of the stone has prevented it assuming its proper place as a commercial marble, and this is its hardness. Owing to this quality the sawing and polishing is costly and therefore it has not come largely into use, beautiful as it is. It is waiting for improved methods of working to be devised, and when this is done the Winooski marble will be in great demand. Like the black marble of Isle La Motte, it has been for a long time manufactured into floor tiling, for which it is admirably adapted. Its beauty and value become more striking, however, when polished and worked up into interior decorations, as may be [illegible] seen in some of the public buildings in the cities and particularly in the capitol at Albany.

Returns from the census will probably give accurate statements of the amount and value of the marble used in our country as well as the sources of supply.

On the authority of Robert W. Welsh, late consul at Carrara, Italy, (Century, June, 1884), we have the information that there is obtained from the four hundred quarries at that place, worked by about five thousand men, an annual product not far from 150,000 tons of marble, this amount saved from the 500,000 tons blasted out of the quarries; of this amount, about 25,000 tons, or one-sixth, comes to the United States.

Our government returns, according to good authority, however, show a considerable increase over the estimated amount, the quantity reaching 35,000 to 40,000 annually, or about 400,000 cubic feet of sawed marble. This pays about 75 cents duty, making the rough blocks worth about $2.50 per cubic foot, or altogether a sum of capital invested in the production of American marble in the States of Vermont, Massachusetts, Connecticut, New York, Pennsylvania, Maryland and Tennessee to be at least $[illegible], two-thirds of this in quarries, one-third in mills and other improvements.

The number of workmen engaged is [illegible], one-half of whom may be considered skilled workmen, engaged in operating and managing machinery.

The amount of annual production is about [illegible] cubic feet, or over [illegible] carloads; and the value, $[illegible]. This is five times the number of cubic feet and three times the total value delivered at our ports, duty paid, of all marble imported per year.

Of this about half is used for building; the other half being used mostly for marble mantels, furniture marble, monuments and cemetery work. This half competes with the imported Italian. Our country now demands for the purposes just named, of the Italian [illegible] cubic feet, of the American [illegible] cubic feet.

The Italian marble does not endure outdoor exposure as well as the better class of American, so that for this first there is no increase of demand for cemetery purposes, while for the same purposes the American marble is rapidly gaining in reputation and use.

Gov. Proctor's estimate of the annual product of the quarries of Vermont is [illegible] cubic feet, with an increase of about [illegible] feet per year. At $[illegible] per cubic foot, $[illegible] may be regarded as a fair estimate of the sum annually realized from the marble quarries of Vermont.

APPENDIX.

LOPMENT OF MACHINERY.

71. Sawing-mill, Charles R. Brewster, West Stockbridge, Mass., 1858.
72. Use of shears for sawing blocks of marble, no date.
73. Channeling machine, George E. Wardwell, Rutland, Vt., 1863.
74. Diamond drill, used in England and France [illegible].
75. Diamond gadders, Sullivan Machine Co., Claremont, N. H.
76. Diamond saw used in [illegible], Conn., 1878.
77. Rock drill, W. L. [illegible], New York, 1869.
78. Various [illegible] arrangements for [illegible] sand and water [illegible], [illegible], [illegible].

[illegible]

[illegible]

[illegible]

[illegible]

[illegible]

[illegible]

[illegible]

[illegible]

EARLY WORK

MARBLE AT MIDDLEBURY.

LOCAL QUESTION OF INVENTION.

[illegible]

[illegible]

[illegible]

Respectfully submitted,

[illegible]
[illegible] Committee.
[illegible]

[illegible]

www.ingramcontent.com/pod-product-compliance
Lightning Source LLC
Chambersburg PA
CBHW022002190326
41519CB00010B/1360

* 9 7 8 3 7 4 2 8 2 8 6 1 3 *